YOUR KNOWLEDGE HAS VALUE

Bibliographic information published by the German National Library:

The German National Library lists this publication in the National Bibliography;
detailed bibliographic data are available on the Internet at http://dnb.dnb.de .

Imprint:

Copyright © 2015 GRIN Verlag, Open Publishing GmbH
Print and binding: Books on Demand GmbH, Norderstedt Germany
ISBN: 978-3-668-03055-8

This book at GRIN:

http://www.grin.com/en/e-book/304401/urea-dtt-soluble-and-insoluble-lens-protein-
in-normal-and-abnormal-human

Ajit Pandya

Urea-DTT soluble and insoluble lens protein in normal and abnormal human eye. A clinical study of cataract

GRIN Publishing

GRIN - Your knowledge has value

Since its foundation in 1998, GRIN has specialized in publishing academic texts by students, college teachers and other academics as e-book and printed book. The website www.grin.com is an ideal platform for presenting term papers, final papers, scientific essays, dissertations and specialist books.

Visit us on the internet:

http://www.grin.com/

http://www.facebook.com/grincom

http://www.twitter.com/grin_com

UREA-DTT SOLUBLE AND INSOLUBLE LENS PROTEIN IN NORAMAL AND ABNORMAL HUMAN EYE-A CLINICAL STUDY OF CATARACT

Dr. Ajit V Pandya

C.U.Shah Science College, Ashram Road, Ahmedabad.

Abstract

More than 5 million cataract surgery is carried out in many third world countries including India. The enzymes associated with the synthesis, catabolism and utilization of glutathione, ascorbic acid and proteins in the lens have been reported with the progression of cataract. It shows increase percentages of urea DTT soluble (UDTTS) and urea DTT insoluble (UDTTIS) fraction in all types of cataractous lenses. The average values of UDTTS and UDTTIS proteins in normal human lenses are 9.21 ± 0.08 and 1.22 ± 0.04 ug/mg (mean $\pm$ s.e.) respectively. Similarly the average values of the same protein fractions in cataractous lenses are 90.23 ± 10.3 and 81.46 ± 7.6 ug/mg (mean $\pm$ s.e.) respectively. The lowest and highest values for UDTTS fraction were obtained in PSC-CS and Brown cataract and are 40.2 and 242.9 ug/mg respectively. The results presented in the current study show both cataract formation and normal ageing are accompanied by a decrease in the solubility of the lens proteins. Most of the color of the lens is associated with urea-insoluble protein fractions. Amongst urea-insoluble fraction, the brown protein fraction was more predominant over yellow protein fraction.

Key words : UDTTS, UDTTIS, Human, Lens, Protein fractions

Introduction

A socio-economic cause of visual defect during aging is not only an important health problem but also one, especially in the tropical countries. Inhibiting or cure in the progression would present a major achievement for human welfare and is therefore one of the priorities of medical research in our country. 41 million people are blind globally of which, 17 million (42%) people are experiencing profound or total loss of vision due to cataract.

Recent research has clearly shown that oxidative stress is a very important risk factor for cataract development (Truscott et al., 2005).

 The lens protein crystalline   account for 80-90% of the soluble proteins of the lens which provide consistency and transparency of the lens (De Jong, 1981). It is suggested by the observation that, at physiological concentration, ascorbic acid inhibits (about 25%) insolubilization and darkening of the lens from exposure to near UV light. Since dehydro ascorbic acid can be reduced by glutathione (Miratashi, 2001) . It seems possible that ascorbic acid could from part of a protective system   as ascorbic acid is antioxidant and in human it is not synthesis in the body but needed through diet (Pandya et al., 2012).

The color of lens is due to glycation and aggregation of lens proteins. Due to aggregation of small molecular weight lens proteins heavy molecular weight aggregates come into existence. The cataractous protein appears to be stabilized by disulphide bonds whereas the protein from normal lenses is not (Dilley, 1975). To find out the changes leading to the production of modified protein fractions from the cataractous and normal human lens and the relationship between them, this study was conducted.

Materials and methods

The urea-soluble, urea-insoluble, Urea- DTT soluble and Urea-DTT insoluble lens proteins fractionation was done by the standard method of Coghland and Augusteyn (1977). Determination of protein was done by the method of Lowry et al., (1951).

The weighed lenses were homogenized in cold distilled water and centrifuged at 4,000 g for 30 mins. and the supernatant was used for other experiment and the residues were dissolved in 8 M urea and centrifuged again. The supernatant obtained was used for other experiment and residues were again dissolved in 8 M urea containing 0.1 nm DTT and centrifuged again. The supernatant obtained was used for the estimation of yellow fraction of lens proteins. The residue were dissolved in 0.3 N NaOH and estimated as brown fractions of lens proteins. The procedure is summarized in figure –1.

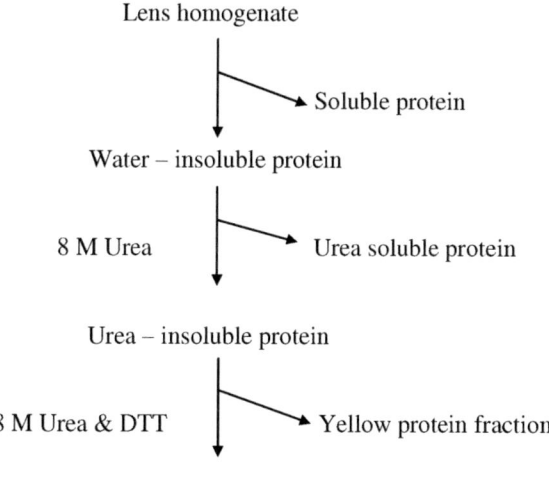

Lens homogenate

Soluble protein

Water – insoluble protein

8 M Urea

Urea soluble protein

Urea – insoluble protein

8 M Urea & DTT

Yellow protein fraction

Brown protein fraction.

Figure - 1

The lenses of the patients who underwent extracapsular cataract surgery   at Nagari Eye Hospital, Ahmedabad, by medical Dr. involved in surgery, were collected. Several eyes with clear lens were obtained from C. H. Samaria eye bank, Red Cross society, Ahmedabad, India.

Statistical analysis: All results were expressed in mean ± s.e.. One way analysis of variance (ANOVA) was used to test the significance of difference and Bonferroni test to test the significance of difference between control and different cataract types. The p value less than 0.05 is considered as significant. The results are expressed graphically by considering values of control lens   as control as 100%.

Results

Table-1 shows the urea DTT soluble (UDTTS) and Urea-DTT insoluble (UDTTIS) fractions of proteins in normal human lens of different age groups studied. It shows highly significant relationship between increase in age and ISP fractions i.e. UDTTS and UDTTIS. The amount of protein in these fractions increases with age. The values in UDTTS and UDTTIS fractions are almost double in last age group (81 - 90 Years) as compared to first age group (31–40 Years). In each age group about 4 to 5 samples were studied for these parameters.

In table-2 the fractionations of cataractous lenses were done and the result obtained is shown. The lowest and highest value for UDITTS fraction was obtained in PSC-CS and Brown cataract and is 40.2 and 242.9 ug/mg respectively. It expresses results of all the fractions in different types of cataracts studied. The lowest and highest values for UDTTIS fraction were obtained in CS and Brown cataract and are 28.6 and 217.5 ug/mg respectively. The values of each fraction in different types of cataracts are significantly different from one another.

The percentages of protein fractions in normal human lenses compared to total proteins are shown in Table-3. But, the percentages of UDTTS and UDTTIS fractions increases with age. Despite of these two findings the linear relationship cannot be established.

Similarly the percentages of UDTTS and UDTTIS fractions are expressed in table-4 for different types of cataracts studied. It shows an increase in percentages of UDITTS and UDTTIS fraction in all types of cataractous lenses.
In Brown cataractous lens the percentages of UDTTS and UDTTIS proteins are 39.32% and 35.22% respectively. In CS type of cataract the percentage of UDTTIS a protein are same and is 18.18%. In CS type of cataract the percentage of UDTTIS protein is lowest i.e. 8.21%. This percentage of protein fractions are obtained considering total proteins as 100%.

In table-5, the average values of protein fractions of normal and cataractous human lenses are shown. The average values of UDTTS and UDTTIS proteins in 28 normal human lenses are 9.21 ± 0.08 and 1.22 ± 0.04 ug/mg (mean ± s.e.) respectively. Similarly the average values of the same protein fractions in 48 cataractous lenses are 90.23 ± 10.3 and 81.46 ± 7.6 ug/mg (mean ± s.e.) respectively. But, at the same time there is an increase in percentage values of UDTTS (879.6%) and UDTTIS (6577.0%) fractions of protein in cataractous lenses compared to normal.

Discussion

The results show both cataract formation and normal ageing are accompanied by a increase in the insolubility of the lens proteins. In this study there was progressive shift of proteins from the soluble to the insoluble fractions as the color of the lens nucleus intensifies. Because of the lowered solubility of the proteins, multiple extractions were required to solubilize protein from both old normal and cataractous lenses (Coghlan and Augustey, 1977). It has been noted by many investigators that the proteins found in nuclear cataracts are similar, in many ways, to the products obtained after photo-oxidation of normal lens protein (Zigman, 1981). Most of the color of the lens is associated with urea-insoluble protein fractions.

It is also clear from the results that because of physical entrapment of soluble proteins in the insoluble proteins in the insoluble fractions. This is especially apparent in Brown lenses where more than 85 % of the protein is insoluble. The lowest insolubility was observed in PSC-CS type of cataract. The remarkable feature of the observation from urea-insoluble fraction was its relationship with the color of lens which is mainly due to inactivation of various enzymes of Glutathion metabolism (Casado, 2001). Many people are particularly interested in the appearance of large amounts of urea-insoluble protein. Amongst urea-insoluble fraction, the brown protein fraction was more predominant over yellow protein fraction. These were produced by fractionation of urea insoluble proteins by extraction with urea containing reducing proteins by extraction with urea containing a reducing agent (DTT) yields a bright yellow soluble protein leaving behind a dark brown residues.

The yellow and brown protein fractions are uniquely associated with different types of cataract especially nuclear cataract and increases with the progression of the cataract (Truscott and Augusteyn, 1977).
The brown color of lens in brown cataract is due to very high amount of brown fraction of protein. It contain about 637 times higher amount of brown fraction compared to yellow fractions of protein. A "heavy molecular weight aggregate" (HMW protein) with an apparent molecular weight in excess of 15×10^6 can be isolated from the water soluble proteins by gel filtration or by differential centrifugation. It has been suggested that this

protein is an intermediate in the insolubilization of lens proteins and that calcium play a key role in this process (Specter and Rothschild, 1973).

Acknowledgement
I am thankful to the institutes – C U SHAH science college, Nagari eye hospital and Dr. Bakulesh Khamar for the support and help during this research work.

References

1. Coghlan, S.D. and Augusteyn, R.C. (1977) : Changes in the distribution of proteins in the aging human lens. Exp. Eye. Res. 25, 603-611.

2. Casado A., de la Torre R., Lopez-Fernandez E. (2001). Antioxidant enzyme levels in red blood cells from cataract patients. *Gerontology*. 47: 186 - 188

3. De Jong, W.W. (1981) : In Molecular and cell biology of the lens (Bloemendal, H ed) Johnely and Sons, N.Y. pp. 221-278.

4. Dilley, K.J. (1975) : The properties of proteins from the normal and cataractous human lens which exists as high molecular weight aggregates in vitro. Exp. Eye. Res. 20, 73-78.

5. Lowery, O.H., Rosebrough, N.J., Farr, A.L. and Randall, R.J., (1951) : Protein measurement with folin-phenol reagent. J. Biol. Chem., 193, 265.

6. Pandya Ajit, (2012) : Comparative Study of Enzymes in Normal and Cataractous   Human Lens, Indian Journal of Advances in Chemical Science 1   (2012) 73-76

7. Miratashi S.A.M., Besharati M.R., Shoja M.R., Rastegar A., Manaviat M.R. (2001). Evaluation of vitamin C concentration of aqueous humour in senile cataract. Med. J. Islamic Acad. Sci. 14: 35 – 40.

8. Spector, A. and Rothschild, C. (1973) : The effect of calcium upon the reaggregation of bovine alpha crystalline. Ivest. Opthalmol. 12, 225-231.

8. Truscott R.J *Age-related nuclear cataract-oxidation is*

the key. Exp Eye Res., 80: 709 -725 (2005).

9. Truscott, R.J.W. and Augusteyn, R.C. (1977). Changes in human lens proteins   during nuclear cataract formation exp. Eye. Re. 24, 159-170.

10. Zigman, S. (1981) : Influence of ageing and light on oxidation : reduction in the lens. In Red Blood Cell and Lens Metabolism. 181-184. Ed, Srivastava, S.K. Publ. Elsevier, North Holland Inc.

TABLE – 1
SOLUBLE AND INSOLUBLE NORMAL HUMAN LENS PROTEINS

AGE (Years)	UDTTS (ug/mg)	UDTTIS (ug/mg)
31 – 40 (5)	5.92 ± 0.05	0.78 ± 0.03
41 – 50 (5)	8.81 ± 0.08	0.85 ± 0.02
51 – 60 (5)	9.26 ± 0.08 *	1.20 ± 0.03 *
61 – 70 (4)	10.35 ± 0.12	1.39 ± 0.09
71 – 80 (5)	9.76 ± 0.10	1.33 ± 0.05
81 –90 (4)	11.31 ± 0.08	1.81 ± 0.03

------► All values are expressed as mean ± S.E.

------► Numbers in the parenthesis are sample sizes.

------► * p-value < 0.05 where as for others p-value < 0.01

------► UDTTS = Urea – DTT Soluble and UDITTS = Urea DTT – Insoluble.

TABLE – 2

UDTTS AND UDTTIS CATARACTOUS HUMAN LENS PROTEINS.

Type of Cataracts	UREA-DTT SOLUBLE (ug/mg)	UREA-DTT INSOUBLE (ug/mg)
NS (4)	123.2 ± 17**	63.4 ± 5*
PSC (4)	106.6 ± 10*	70.0 ± 5*
CS (4)	78.4 ± 8*	28.6 ± 4*
NS, PSC (5)	54.4 ± 6*	73.4 ± 7*
PSC, CS (6)	40.2 ± 5*	44.2 ± 5**
NS, CS (6)	52.4 ± 7*	57.6 ± 5*
NS, PSC, CS (6)	97.6 ± 15**	106.3 ± 8*
CS, NS, PP (3)	47.1 ± 5*	62.7 ± 8**
MATURE (5)	59.5 ± 5*	90.9 ± 8*
BROWN (5)	242.9 ± 25*	217.5 ± 20*

------▶ All values are expressed as mean ± S.E.
------▶ Number in the parenthesis are sample sizes.
------▶ p-value * < 0.01, P-value ** < 0.05.

TABLE – 3
PERCENTAGES OF NORMAL HUMAN LENS PROTEINS FRACTIONS COMPARED TO TOTAL PROTEIN (100%)

AGE (Years)	UREA-DTT SOLUBLE	UREA-DTT INSOUBLE
31 – 40 (5)	1.62	0.21
41 – 50 (5)	2.41	0.23
51 – 60 (5)	2.51	0.32
61 – 70 (4)	2.58	0.34
71 – 80 (5)	2.48	0.33
81 –90 (4)	2.78	0.44

------▶ Numbers in the parenthesis are sample sizes.
----------→ Total protein is 100 %

TABLE – 4

PERCENTAGES OF PROTEIN FRACTIONS IN CATARACTOUS
HUMAN LENS COMPARED TO TOTAL PROTEIN

UREA-DTT SOLUBLE	Type of Cataracts	Number of samples	UREA-DTT INSOUBLE
29.61	NS	5	15.25
25.22	PSC	5	16.55
22.52	CS	4	08.21
14.15	NS, PSC	6	19.09
12.11	PSC, CS	5	13.32
14.55	NS, CS	6	15.99
23.53	NS, PSC, CS	6	25.62
11.85	CS, NS, PP	4	15.77
18.18	MATURE	4	27.79
39.32	BROWN	5	35.22

-----------→ Total protein is 100 %

TABLE – 5

UDTTS AND UDTTIS PROTEIN FRACTIONS
OF NORMAL AND CATARACTOUS HUMAN LENSES

TYPE	UDTTS (ug/mg)	UDTTIS (ug/mg)
NORMAL (n = 28) 60 ± 10 Yrs.	9.21 ± 0.08	1.22 ± 0.04
CATRACTOUS (n = 40) 58 ± 12 Yrs.	90.23 ± 10.3	81.46 ± 7.6

-------► All values are expressed as mean ± S.E.
-------► n = sample sizes.
-------► For all p-value < 0.05